FLORA OF TROPICAL EAST AFRICA

ALISMATACEAE

Susan Carter

Perennial, rarely annual, aquatic, swamp or marsh herbs, lactiferous. Rhizome very short; roots short, fibrous. Leaves erect, rarely floating or submerged, basal; petiole with an expanded, sheathing base; leaf-blade entire, linear-lanceolate to ovate, with a decurrent to sagittate base, acute to rounded apex. Inflorescence compound or simple, of whorls of branches or flowers, rarely pseudo-umbellate or with solitary flowers ; bracts 2 or 3 at the base of each whorl, and sometimes several bracteoles. Flowers regular, bisexual or unisexual. Sepals 3, persistent, herbaceous. Petals 3, deciduous, rarely 0. Stamens 3, 6, 9 or more; filaments filiform or flattened; anthers 2-celled, dehiscing longitudinally and laterally. Carpels superior, free or joined at the base, 3–∞, in a whorl or spiral, unilocular; style terminal or ventral; ovules 1, 2 or many, basal and erect, or situated on the ventral suture. Fruit indehiscent. Seeds oblong, indented laterally to follow the form of the horseshoe-shaped embryo, smooth, wrinkled or ridged, without endosperm.

In common with most water-plants members of this family show great morphological variability, especially in the size and shape of the leaves. This has led to much confusion in that extreme forms of a species have often been described as distinct species or varieties. Such extremes are prevalent under tropical conditions, notably the more robust forms. However, among specimens examined from the tropical African areas, there was always a good proportion of the " typical " forms which, in most cases, have been described from European or Asian material.

Inflorescence pseudo-umbellate of 1 whorl of 1 to 3
    flowers .   .   .   .   .   .   1. **Ranalisma**
Inflorescence compound or of more than 2 whorls :
  Inflorescence unbranched ; flowers sessile or sub-
    sessile ; monoecious   .   .   .   6. **Wisneria**
  Inflorescence branched, or if unbranched then flowers
    pedicellate :
    Leaves sagittate or deeply cordate :
      Achenes ∞, much compressed laterally, spirally
        arranged on a convex receptacle ; monoe-
        cious   .   .   .   .   .   2. **Sagittaria**
      Achenes 5–30, not laterally compressed, bunched
        on a flat or small receptacle ; flowers
        hermaphrodite or polygamous :
        Leaves deeply cordate ; flowers herma-
          phrodite   .   .   .   .   4. **Caldesia**
        Leaves sagittate; flowers polygamous . . 5. **Lymnophyton**
    Leaves lanceolate to ovate; base cuneate to trun-
      cate :
      Achenes in a whorl ; lateral faces smooth ;
        flowers and branches in whorls of at least 4 ;
        flowers hermaphrodite   .   .   .   3. **Alisma**

1

Achenes bunched ; lateral faces with a flange ;
flowers and branches usually in whorls of 3 ;
dioecious .   .   .   . 7. **Burnatia**

1. RANALISMA

Stapf in Hook., Ic. Pl. 27 : t. 2652 (1900)

[*Echinodorus* sensu Wright in F.T.A. 8 : 211 (1901); sensu Buchen. in E.P.
IV. 15 : 23 (1903), pro parte, non L. C. Rich.]

Perennial stoloniferous glabrous marsh herbs. Leaves erect, or spreading ;
leaf-blade ovate to linear-lanceolate ; base cordate to cuneate. Inflorescence
pseudo-umbellate, 1–3-flowered with 2 membranous bracts ; flowers sometimes replaced by vegetative buds—the peduncle bends over stolon-like and

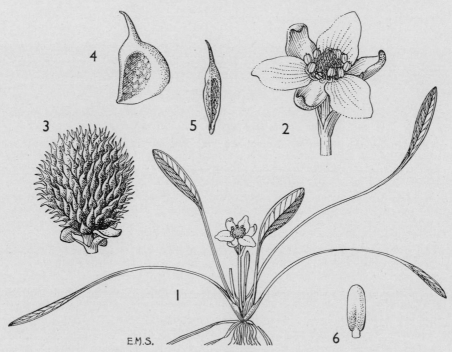

FIG. 1.   *RANALISMA HUMILE*—**1,** plant, × 1 ; **2,** flower, × 3 ; **3,** fruit, × 3 ; **4,** achene, lateral view,
× 6 ; **5,** achene, dorsal view, × 6 ; **6,** seed, × 6. 1, 2 from *Richards* 6380 ; 3–6 from *Milne-Redhead* 5027.

the buds develop into new plants. Flowers bisexual. Sepals 3, reflexed in
the fruiting stage. Petals 3, larger than the sepals, white. Stamens
(6–)9(–12) ; filaments filiform. Carpels ∞, free on a convex receptacle ;
styles terminal, hooked ; ovules solitary, basal. Fruit a globose head of
achenes spirally arranged on the elongated receptacle ; achenes strongly
compressed laterally, glandular, winged, with a terminal beak formed by the
remains of the style.

A genus of 2 species, closely allied to *Echinodorus* L. C. Rich. of which there are no
representatives in Africa. The typical species, *R. rostratum* Stapf, is confined to tropical
SE. Asia and the Malay Peninsula.

R. humile (*Kunth*) *Hutch.* in F.W.T.A. 2 : 303, fig. 280 (1936). Type :
Senegambia [Senegal], nr. Richard-Tol, *Lelièvre* (B, holo.†)

A small herb 2–4 cm. high, which may reach up to 15 cm. if growing in
water. Stolons short. Petiole 1–10 cm. long, flattened ; leaf-blade ovate

to ovate-lanceolate, 0·5–1·5 × 1–3 cm. when emersed, or linear-lanceolate, up to 0·5 × 10 cm. when submerged; apex acute; base rounded to cuneate or slightly decurrent; 3, rarely 5 nerves radiating from the apex of the petiole. Inflorescence of 1 flower, rarely 2; peduncles equal to the petioles; bracts 3–5 mm. long; vegetative buds sometimes present. Sepals ovate, 4 × 3 mm. Petals 6 × 4 mm., very delicate. Stamens 8–12; filaments 2·5 mm. long; anthers ovate, less than 1 mm. long. Carpels obovoid, compressed; styles 1 mm. long. Achenes densely packed in a buff-coloured, globose head 6–10 mm. in diameter, obliquely obovoid, 2–2·5 × 1·5–2 mm., the remains of the style rigid, 1 mm. long. Fig. 1.

TANGANYIKA. Tabora District : Ugalla River, 192 km. SW. of Tabora, 1 Oct. 1948, *Bally* 7527 !
DISTR. **T4**; known in East Africa from one gathering only, but is widespread from Senegal to the Sudan, and through the Belgian Congo to Northern Rhodesia.
HAB. Marshy places near rivers ; 1140 m.

SYN. *Alisma humile* Kunth, Enum. Pl. 3 : 154 (1841)
 Echinodorus humilis (Kunth) Buchen. in Pringsh. Jahrb. 7 : 28 (1868) ; F.T.A. 8 : 211 (1901) ; E.P. IV. 15 : 26 (1903)
 Sagittaria humilis (Kunth) O. Ktze., Rev. Gen. 3 : 326 (1891)

2. SAGITTARIA

L., Sp. Pl.: 993 (1753) & Gen. Pl., ed. 5 : 429 (1754) ; C. Bogin in Mem. New York Bot. Gard. 9 : 179 (1955)

Perennial, occasionally annual, monoecious, rarely dioecious, marsh or aquatic herbs. Leaves erect, floating or submerged; leaf-blade of the erect leaves linear to broadly ovate, often with sagittate base; that of the floating leaves linear to ovate, with cuneate to cordate base ; submerged leaves rarely lanceolate, usually modified into linear phyllodes, occasionally terete and spongy. Inflorescence erect or sometimes floating, with 1–12 whorls, each usually 3-flowered, the upper ♂, the lower ♀ ; flowers occasionally replaced by branches ; bracts 3 ; pedicels often thickened and recurved in fruit. Sepals 3, reflexed in the ♂ flowers, appressed, spreading or reflexed in the ♀ flowers. Petals 3, white, rarely pink, sometimes with a purple spot near the base. Stamens 9–∞ ; filaments filiform, sometimes flattened or broadened at the base. Carpels ∞, free, spirally arranged on a convex receptacle ; style ventral ; ovules solitary, basal. Fruit a globose head of achenes which are much compressed laterally, winged and beaked.

A genus of about 20 species, 3 of which occur in the Old World. Of these *S. guyanensis* H.B.K. is the only one native in Africa, and has so far been found in Gambia, Nigeria, the Sudan and Madagascar. Of all the *Sagittaria* species, *S. montevidensis* Cham. & Schlecht. subsp. *montevidensis* is the only widely cultivated one and, as in tropical East Africa, it has sometimes escaped and become established.

S. montevidensis *Cham. & Schlecht.* in Linnaea 2 : 156 (1827). Type : Brazil, Rio Grande do Sul, *Sellow* (B, holo.†)

Annual or perennial aquatic herb. Leaves erect, floating or submerged ; blade of erect leaves linear to broadly ovate, 0·5–25 × 2–20 cm., sagittate, with the lobes up to 20 cm. long, rarely reduced or absent; blade of floating leaves ovate, up to 1·5 × 2·5 cm.; submerged leaves reduced to phyllodes, up to 45 cm. long. Peduncle erect, 7–75 cm. high, with 2–12 whorls of flowers, the lowest sometimes of branches ; bracts connate at their bases, up to 2·5 cm. long, membranous or sometimes thickened ; pedicels of ♂ flowers 1–4 cm. long ; pedicels of ♀ flowers thickened, recurved, 1–7 cm. long. Sepals ovate, 0·6–1·5 × 0·6–1·5 cm., appressed in the female flowers. Petals larger, up to 2·2 cm. long, white, sometimes with a purple mark at the base. Stamens of ♂ flowers 12–∞ ; filaments linear or flattened, glabrous or pubescent, 0·7–3·5 mm. long; anthers oblong, 0·4–1·3 mm. long; ♀ flowers occ-

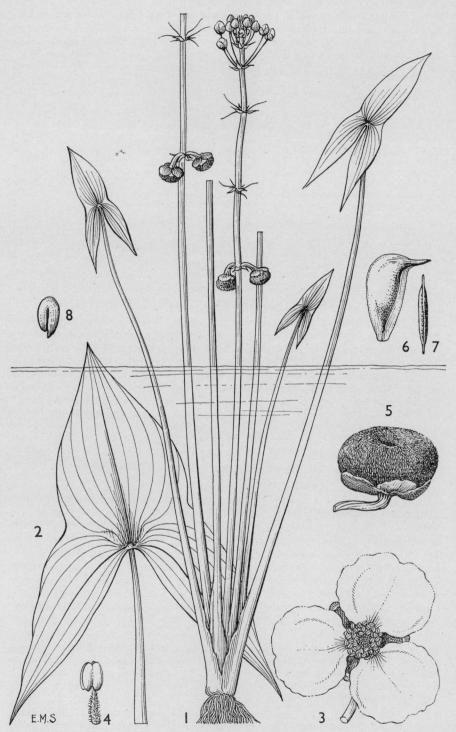

E.M.S

FIG. 2. *SAGITTARIA MONTEVIDENSIS* subsp. *MONTEVIDENSIS*—**1**, plant, × ⅛ ; **2**, leaf, × ⅔ ; **3**, ♂ flower, × 1½ ; **4**, stamen, × 4 ; **5**, ♀ flower in fruit, × 1½ ; **6**, achene, lateral view, × 9 ; **7**, achene, dorsal view, × 9 ; **8**, seed, × 9. From *Verdcourt* 78.

asionally with a ring of 9–12 functional stamens. Abortive carpels in ♂ flowers about as many as stamens, 1·5 × 0·5 mm. ; carpels in ♀ flowers ∞, 2 × 1 mm. Fruiting head up to 2·5 cm. in diameter; achenes up to 1·5 × 3 mm. with a beak up to 1 mm. long.

subsp. **montevidensis**

Perennial. Leaves typically erect and sagittate, rarely floating or submerged. Peduncle with 3–12 whorls of flowers or branches ; ♂ flowers in whorls of 3–6 with 3–6 bracts ; ♀ flowers in whorls of 3 with 3 membranous bracts. Petals with a purple mark at the base. Stamens 20–30 in ♂ flowers ; filaments linear, 1·8–3·5 mm. long, sparsely pubescent ; functional stamens rarely present in ♀ flowers. Fig. 2.

TANGANYIKA. Lushoto District : Nderema, 1 Aug. 1913, *Grote* 5626 ! & 31 Oct. 1935, *Greenway* 4152 ! & 18 Feb. 1950, *Verdcourt* 78 !

DISTR. **T3** ; warm temperate South America east of the Andes, and northward to coastal Ecuador ; introduced and naturalized in the E. Usambara Mts. around Amani and Nderema, but so far not found elsewhere in tropical Africa.

HAB. Ponds, swamps, and at the edges of slow-flowing streams ; 800–900 m.

SYN. *S. montevidensis* sensu Buchen. in E.P. IV. 15 : 43 (1903)

According to Bogin there are three other subspecies—subsp. *calycina* (Engelm.) Bogin and subsp. *spongiosa* (Engelm.) Bogin both from North America, and subsp. *chilensis* (Cham. & Schlecht.) Bogin from temperatute Chile.

3. ALISMA

L., Sp. Pl. : 342 (1753) & Gen. Pl., ed. 5 : 160 (1754)

Perennial aquatic or marsh herbs. Leaves erect ; leaf-blade linear-lanceolate to ovate ; apex acute ; base decurrent to cordate. Inflorescence compound, pyramidal ; flowers and branches in whorls of 4 or more, each whorl subtended by 3 bracts and several smaller bracteoles. Flowers bisexual. Sepals 3, appressed in the fruiting state. Petals 3, larger than the sepals. Stamens 6 ; filaments filiform. Carpels many, free ; style ventral, erect or curved ; stigma papillose ; ovules solitary, basal. Achenes many, in a whorl on the flat receptacle, compressed laterally, ridged dorsally.

A world-wide genus of several species and sub-species.

A. plantago-aquatica *L.*, Sp. Pl. : 342 (1753) ; F.T.A. 8 : 207 (1902) ; E.P. IV. 15 : 13 (1903), pro parte. Type : Europe (LINN, holo. !)

Aquatic herb. Petiole (10–)15–30(–40) cm. long ; leaf-blade ovate, (5–)10–15(–25) × (2·5–)5–7(–12) cm.; base rounded to subcordate; nerves 5–9, not all radiating from the apex of the petiole. Peduncle (2–)6–10 dm. high. Inflorescence of about 6 whorls of branches and flowers ; bracts membranous, lanceolate, those of the lowest whorl 1–2 × 0·5 cm., with several smaller bracteoles. Pedicels up to 2 cm. long. Sepals ovate, 2 mm. long. Petals 4 mm. long, white, delicate. Stamens 2 mm. long; anthers 0·75 mm. long, narrow; filaments attached above the base. Carpels ovate, compressed, less than 1 mm. long; style about 1 mm. long, erect or slightly curved. Achenes broadly ovate, with 2 or 3 ridges on the dorsal surface, pale brown to yellowish. Fig. 3, p. 6.

UGANDA. Toro District : Bunyangabu, *Snowden* 103 !

KENYA. Trans-Nzoia District : NE. Elgon, Mar. 1949, *Tweedie* 730 ! ; N. Nyeri District: Nyeri, May 1956, *D. F. Smith* ! ; Nairobi District : Karura forest outskirts, 21 Feb. 1953, *Verdcourt & Steele* 914 !

TANGANYIKA. Mbulu District : Mbulu, Oct. 1925, *Haarer* 98B ! ; Arusha District : Nduruma–Arusha road, Dec. 1927, *Haarer* 927 ! ; Moshi District : 8 km. W. of Moshi by R. Njoro, 3 Nov. 1955, *Milne-Redhead & Taylor* 7036 !

DISTR. **U2** ; **K3, 4** ; **T2** ; temperate regions of Europe, Asia, Australia and Africa ; according to Samuelsson (Arkiv Bot. 24A, No. 7 (1932)) distribution in the North American continent is limited to subsp. *brevipes* (Greene) Sam., while subsp. *orientale* Sam. is found in eastern and central Asia

HAB. Marshes and river margins, usually growing in shallow water ; 900–2280 m.

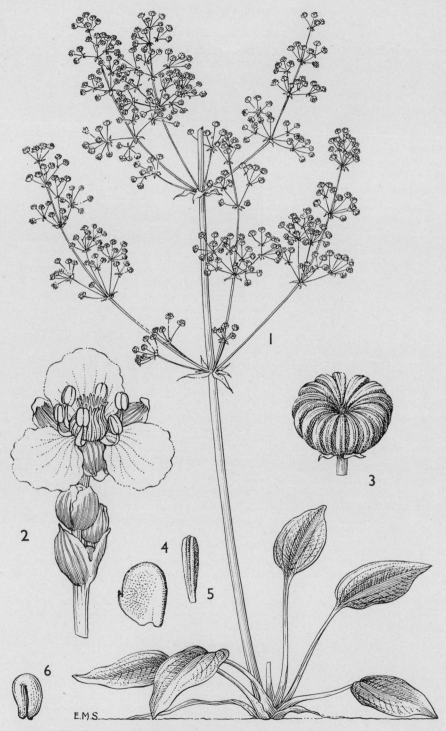

FIG. 3. *ALISMA PLANTAGO–AQUATICA*—**1,** plant, × ½ ; **2,** part of inflorescence, × 6 ; **3,** fruit, × 9 ; **4,** achene, lateral view, × 12 ; **5,** achene, dorsal view, × 12 ; **6,** embryo, × 12. 1, 2 from *Milne-Redhead & Taylor* 7036 ; 3–6 from *D. F. Smith.*

4. CALDESIA

Parl., Fl. Ital. 3 : 599 (1858)

Alisma sect. *Caldesia* (Parl.) Hook. f. in G.P. 3 : 1005 (1883)

[*Alisma* sensu C. H. Wright in F.T.A. 8 : 207 (1902), pro parte, non L.]

Perennial or annual aquatic and marsh herbs. Leaves floating ; leaf-blade broadly elliptic to broadly ovate ; apex rounded to obtuse ; base truncate to deeply cordate. Peduncle longer than the leaves ; inflorescence compound, pyramidal, consisting of whorls of 3 branches or 3 flowers subtended by 3 lanceolate bracts, and 2 or 3 smaller bracteoles. Flowers bisexual. Sepals 3, spreading or reflexed. Petals 3, white. Stamens 6(–11) ; filaments filiform or flattened ; anthers oblong. Carpels 2–9(–20), free, crowded on a small receptacle ; style ventral ; ovules solitary, basal. Achenes swollen, smooth, ridged or warty ; style persistent, ventral.

A genus which is confined to the Old World. There are 4 species of which only one, *C. parnassifolia* (Bassi) Parl. does not have a tropical distribution.

C. reniformis (*D. Don*) *Makino* in Bot. Mag. Tokyo 20 : 34 (1906). Type : Nepal, *Wallich* (K, ? iso. !)

Leaf-blade broadly ovate, 5–9 × 5·5–12 cm. ; apex rounded or broadly obtuse ; base deeply cordate; ' nerves 13–17, radiating from the apex of the petiole. Peduncle 4–7 dm. high; inflorescence with 4–8 whorls, the lowest always of branches ; bracts oblong-lanceolate, the lowest about 1 cm. long, thickened ; bracteoles much shorter, membranous ; flowers sometimes replaced by vegetative buds ; pedicels 1–4 cm. long. Sepals spreading, broadly ovate, 4 × 3 mm. Petals delicate, a little larger than the sepals. Stamens 6 ; filaments 1·5–2 mm. long, flattened and broadened at the base; anthers broad, nearly 1 mm. long. Carpels 10–15, ovate, 1 mm. long ; style slightly hooked, 1 mm. long. Achenes about 8, with a spongy exocarp (not usually evident in dried material) and hard endocarp with 5–9 longitudinal ridges, 3 mm. long. Fig. 4, p. 8.

UGANDA. Mengo District : Nambigiluwa swamp, Entebbe, Nov. 1932, *Eggeling* 700! ; King's Lake, Kampala, Jan. 1936, *Hancock & Chandler* 136 !

TANGANYIKA. Arusha District : Lake Balbal [Duluti], *Uhlig* 512 †

DISTR. U4 ; T2 ; widespread, but not common, from India, Malaysia, China and Japan to Australia, Madagascar, tropical and North Africa

HAB. Swamps, shallow lakes and river-margins ; 1125–1170 m.

SYN. *Alisma reniforme* D. Don, Fl. Nepal : 22 (1825)
     A. parnassifolium Bassi var. *majus* Micheli in DC., Mon. Phan. 3 : 36 (1881). Type : Nepal, *Wallich* (K, ? iso. !)
     Caldesia parnassifolia (Bassi) Parl. var. *major* (Micheli) Buchen. in E.P. IV. 15 : 16 (1903)
     C. parnassifolia (Bassi) Parl. var. *nilotica* Buchen. in E.P. IV. 15 : 16 (1903). Type : Sudan, Bahr el Ghazal region, *Schweinfurth* 1166 and ser. III. 222 (K, syn. !)

NOTE. This species is closely related to the European *C. parnassifolia* (Bassi) Parl., which is smaller in all its parts, has a simpler inflorescence and a leaf with an obtuse apex and fewer nerves.

5. LIMNOPHYTON

Miq., Fl. Ind. Bat. 3 : 242 (1855)

Perennial aquatic and marsh herbs. Leaves erect; leaf-blade with obtuse to rounded apex ; base sagittate or cuneate ; blade and petiole glabrous, scabrid or pubescent. Peduncle longer than the leaves ; inflorescence of 4–7 whorls of flowers, the lowest sometimes with 2–3 branches ; upper whorls of ♂ flowers, lower of ♀ and ♂ flowers ; bracts 3, membranous, glabrous or pubescent ; bracteoles 3 or more, glabrous or pubescent, smaller than

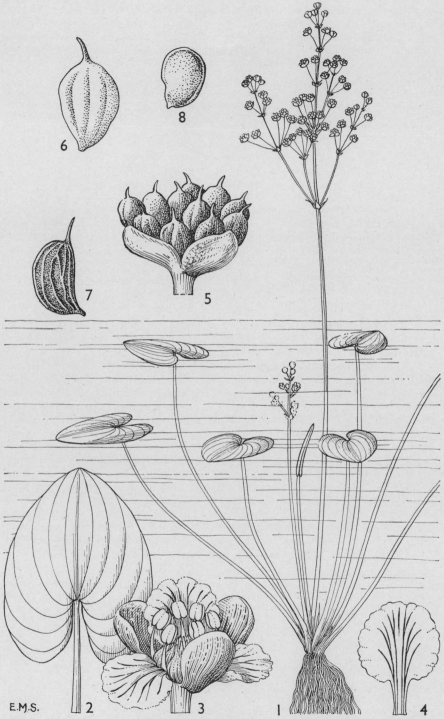

FIG. 4.   *CALDESIA RENIFORMIS*—**1,** plant, × ⅓ ; **2,** leaf, × ⅔ ; **3,** flower, × 6 ; **4,** petal, × 6 ; **5,** fruit,
× 3 ; **6,** achene, dorsal view, with spongy exocarp evident, × 6 ; **7,** achene, lateral view, exocarp not
evident, × 6 ; **8,** seed, × 6. 1, 2, 7 from *Hancock & Chandler* 136 ; 3, 4 from *Knowles* 44 ; 5, 6, 8 from
Schweinfurth ser. III, 222.

the bracts. Sepals 3, spreading. Petals 3, larger than the sepals, white.
Stamens 6 ; filaments broadened at the base. Carpels 10–30, free, ovoid ;
style ventral ; ovules solitary, basal ; abortive carpels in the ♂ flowers 0.
Achenes crowded on the small receptacle, swollen, with lateral air-chambers,
ridged.

A genus which is regarded by some authors as being monospecific. However, the diff-
erences between the two species described here are constant, and too diverse to be
variations of the same species. A third species, *L. fluitans* Graebn., with linear-lanceolate
leaves and small (3 × 2 mm.) achenes which are sharply ridged and shiny when dried,
has been found in the French Cameroons and S. Nigeria. This appears to be very rare,
and has not so far been discovered in our area.

Bracts triangular, glabrous ; bracteoles 3–6 ; flowers
5–10 in each whorl ; pedicels of fruiting flowers
thickened, 1–2 mm. in diameter ; achenes reticulately ridged, pale brown, rough-textured ;
leaves glabrous ; angle between the lobes more
than 90° . . . . 1. *L. obtusifolium*
Bracts lanceolate, pubescent ; bracteoles 10 or more ;
flowers 15 or more in each whorl ; pedicels of
fruiting flowers slender, 0·5 mm. in diameter ;
achenes longitudinally ridged when dried, red-
brown, smooth and glossy ; leaves pubescent ;
angle between the lobes less than 90° . . . 2. *L. angolense*

1. **L. obtusifolium** (*L.*) *Miq.*, Fl. Ind. Bat. 3 : 242 (1855) ; F.T.A. 8 :
209 (1902) ; E.P. IV. 15 : 22 (1903). Type : Asia, *Plukenet* t. 220, fig. 7
in Herb. Sloane, vol. 97, fol. 181 (BM, lecto. !)

Petioles 25–50 cm., triangular in section ; leaf-blade obovate in the
seedling stage (which sometimes flowers and bears fruit) 3·5 × 1·5 cm. ;
mature leaves sagittate, 6–9 cm. measured along the midrib, 8–13 cm. wide
on a level with the petiole apex ; lobes 7–11 cm. long with the angle between
them more than 90° ; nerves 13–17 ; blade and petiole glabrous, rarely
slightly pubescent or scabrid at the petiole apex. Peduncle as long as the
petioles ; inflorescence of 5–7 whorls ; bracts glabrous, triangular, 1–2 ×
1 cm., with rounded apex ; bracteoles 3–6, glabrous, about 3 mm. long ;
basal whorl with 6–10 ⚥ flowers, 0–2 ♂ flowers ; pedicels 2–4 cm. long,
slender in ♂ flowers, but thickened to 1–2 mm. in diameter and much recurved in fruit. Sepals about 6 × 3·5 mm. in ⚥ flowers, 4 × 2 mm. in
♂ flowers. Petals delicate. Stamens 6 ; filaments 2 mm. long ; anthers
oblong, 1 mm. long ; stamens much smaller in the ⚥ flowers. Carpels 2 mm.
long ; style 0·5 mm. long. Achenes ovoid, 4 × 3 mm., shortly stalked,
reticulately ridged, rough-textured, pale brown. Fig. 5/1–5, p. 10.

UGANDA. Teso District : Soroti, near Wera Swamp, 15 Sept. 1954, *Lind* 332 !
KENYA. Northern Frontier District : Tanaland, Kolbio, 29 Aug. 1945, *Mrs. J. Adam-
son* 120 *in Bally* 6020 ! ; Kilifi District : Arabuko, Oct. 1929, *R. M. Graham* 2164 ! ;
Lamu District : coastland near Lamu, Dec. 1875, *Hildebrandt* 1912 !
TANGANYIKA. Shinyanga District : Seseku, 13 June 1931, *B. D. Burtt* 3315 ! ; Lushoto
District : Mkomazi, 23 April 1934, *Greenway* 3969 ! ; Uzaramo District : 19 km. W.
of Dar es Salaam, 7 July 1957, *Welch* 375 !
DISTR. U3 ; K1, 7 ; T1, 3, 6 ; widespread from tropical Africa and Madagascar to
India, Ceylon and the Malay Islands
HAB. Swamps and pools in standing or slow-flowing water ; 0–1140 m.

SYN. *Sagittaria obtusifolia* L., Sp. Pl. : 993 (1753)
    Alisma sagittifolium Willd., Sp. Pl. 2 : 277 (1799). Type : Guinea, (collector
        unknown No. 7107 in B–W, holo., K, photo. !)
    A. kotschyi Hochst. ex Al. Br. in Flora 26 : 499 (1843). Type : Sudan, Kordofan,
        Arashkol, *Kotschy* 169 (BM, K, iso. !)
    A. obtusifolium (L.) Thwaites, Enum. Pl. Zeyl.: 332 (1864)
    Dipseudochorion sagittifolium (Willd.) Buchen. in Flora 48 : 241 (1865)

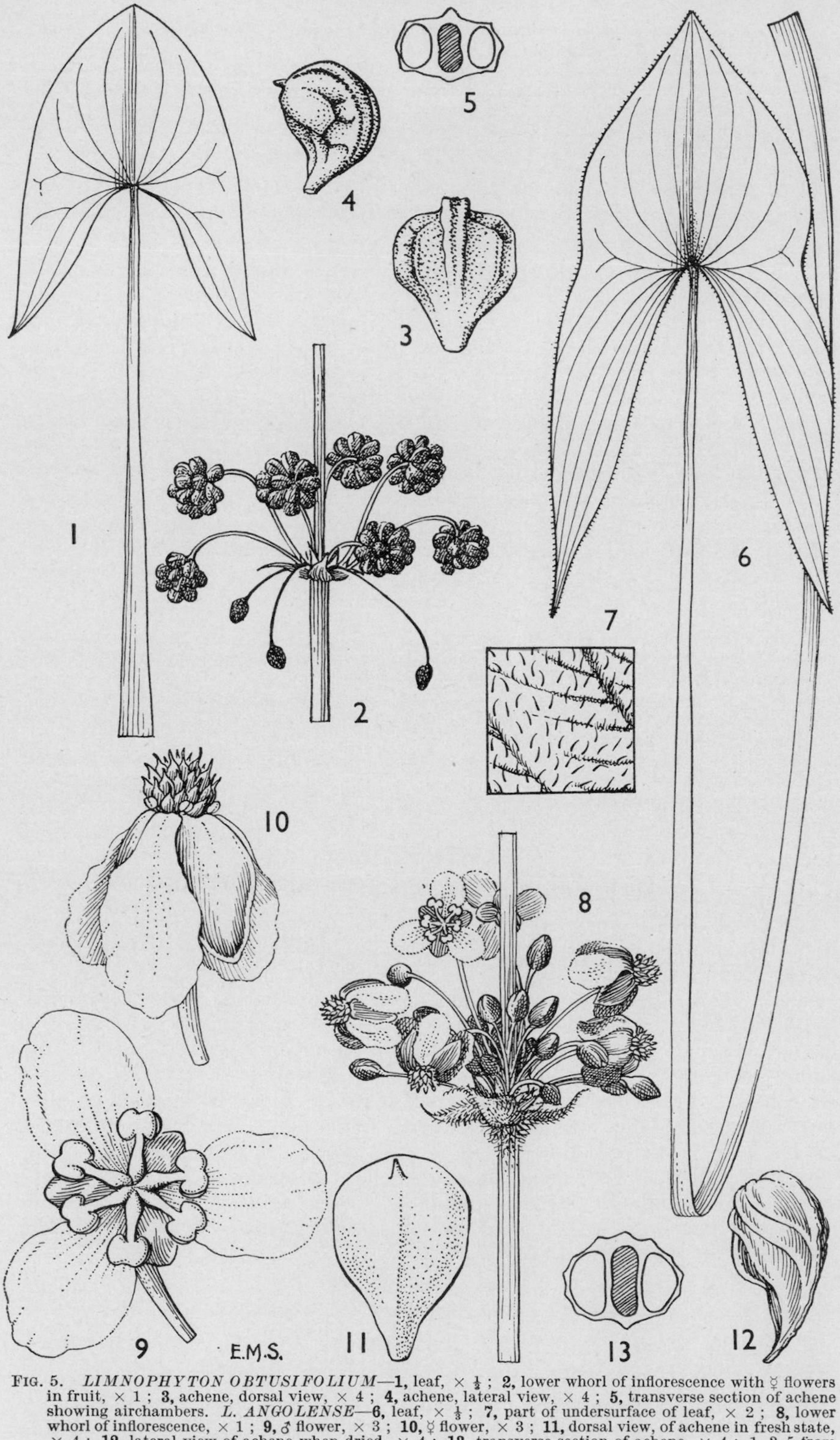

FIG. 5. *LIMNOPHYTON OBTUSIFOLIUM*—1, leaf, × ½ ; 2, lower whorl of inflorescence with ☿ flowers in fruit, × 1 ; 3, achene, dorsal view, × 4 ; 4, achene, lateral view, × 4 ; 5, transverse section of achene showing airchambers. *L. ANGOLENSE*—6, leaf, × ⅓ ; 7, part of undersurface of leaf, × 2 ; 8, lower whorl of inflorescence, × 1 ; 9, ♂ flower, × 3 ; 10, ☿ flower, × 3 ; 11, dorsal view, of achene in fresh state, × 4 ; 12, lateral view of achene when dried, × 4 ; 13, transverse section of achene, × 4 ; 1, 3–5 from *Wailly* 5387 ; 2 from *Welch* 375 ; 6–8 from *Brown* 136 ; 9, 11 from *Milne-Redhead* 4082   10 from *Eggeling* 865 ; 12, 13 from *Michelmore* 315.

Caldesia sagittarioïdes Ostenfeld in Philipp. Journ. Sc. 9 : 259 (1914). Type : Annam, *C. B. Robinson* 1168 (C, holo.)

Limnophyton obtusifolium (L.) Miq. var. *lunatum* Peter, F.D.O.-A. : 119 & Anhang 10 (1929). Type : Tanganyika, Uzaramo, Dar es Salaam, *Peter* 44670 (B, holo.†)

L. parvifolium Peter, F.D.O.-A. : 119 and Anhang 10 (1929). Type : Tanganyika, Uzaramo, Dar es Salaam, *Peter* 46897 (B, holo. †)

2. **L. angolense** *Buchen.* in E.P. IV. 15 : 23 (1903). Types : Angola, Malange, *Mechow* 281 (B, holo.†) ; and Kubango, *Baum* 364 (BM, K, isosyn. !)

Petioles about 85 cm. long ; blade of mature leaves sagittate, usually constricted laterally on a level with the petiole apex, 10–15 cm. measured along the midrib, 10–20 cm. wide at the constriction ; lobes 13–20 cm. long, with the angle between them less than 90° ; nerves 17–25 ; petiole and blade (especially on the lower surface) pubescent, the older, larger leaves rarely glabrous. Inflorescence of 4–6 whorls ; bracts lanceolate, acute, 2–2·5 × 0·5 cm., pubescent on the back and margins ; bracteoles 10–20, pubescent, about 1 cm. long ; basal whorl with 7–11 ⚥ flowers and 8–12 ♂ flowers ; pedicels 2–4 cm. long, not more than 0·5 mm. in diameter in the fruiting flowers and weakly recurved. Flower measurements as for *L. obtusifolium* (L.) Miq. Achenes ovoid, up to 5 × 4 mm. when dried, with the stalk 1·5 mm. long, longitudinally ridged, smooth and glossy red-brown; when fresh they are up to 8 × 5 mm., with the ridges scarcely evident. Fig. 5/6–13.

UGANDA. Lango District : Kachung, *Mukasa in Eggeling* 3951 ! ; Masaka District : Lake Nabugabo, 6 Oct. 1953, *Drummond & Hemsley* 4650 ! ; Mengo District : Namanve Swamp, Kiagwe, Aug. 1932, *Eggeling* 865 !

TANGANYIKA. Bukoba District : Buyango, *Gillman* 297 ! ; Mwanza District : Ihelele, Mbarika, 3 Nov. 1952, *Tanner* 1117 ! ; Kigoma District : Kitolo, 22·5 km. N. of Nguruka, 7 Oct. 1949, *Bally* 7550 !

DISTR. U1 4 ; T1, 4 ; scattered distribution from Sierra Leone, Liberia and Nigeria, to eastern Africa and Northern Rhodesia

HAB. Swamps ; 1000–1200 m.

6. WISNERIA

Micheli in DC., Mon. Phan. 3 : 82 (1881)

Perennial aquatic herbs, floating or loosely anchored. Leaves submerged or partly emersed ; petioles up to 60 cm. depending on the depth of the water, terete ; leaf-blade linear to linear-lanceolate, much shorter than the petioles. Peduncles shorter than the leaves ; inflorescence emersed, spicate, of unbranched whorls of sessile or subsessile flowers; upper whorls of ♂ flowers, lower of ♀ flowers ; vegetative buds sometimes present in the lower whorls ; bracts 3, connate. Male flowers : 4–8 flowers in each whorl, shortly pedicellate ; bracteoles ovate ; sepals 3, ovate ; petals 0–3, much smaller than the sepals, membranous ; stamens 3, filaments flattened ; abortive carpels 2 or 3. Female flowers : 3 to each whorl, ebracteolate ; sepals 3, ovate ; petals 0–3, very small, membranous ; staminodes 3 ; carpels 3–6, free ; styles terminal ; stigmas papillose, ovules solitary, basal ; achenes ovate, swollen, with lateral air-chambers when mature, beaked with the remains of the style, smooth or ridged.

A genus of 3 species all of which appear to be rare. The type species, *W. triandra* (Dalzell) Micheli, occurs in eastern India, the other two in tropical Africa and Madagascar.

W. filifolia *Hook f.* in G.P. 3 : 1007 (1883) ; E.P. IV. 15 : 61 (1903) ; K.B. 12 : 333 (1957). Type : Central Madagascar, *Baron* 591 (K, holo. !)

Roots sometimes anchored in floating weeds. Petioles submerged, very long, 10–60 cm. ; leaf-blade partly emersed, linear, scarcely distinguishable

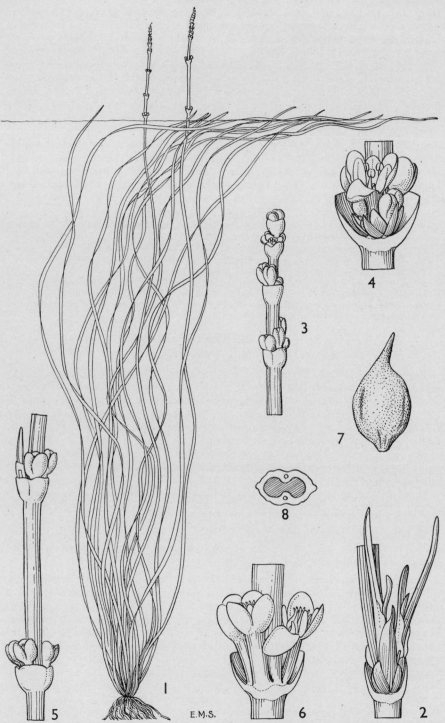

FIG. 6. *WISNERIA FILIFOLIA*—**1,** plant, × ⅓ ; **2,** vegetative buds from a lower whorl of inflorescence,
× 3 ; **3,** ♂ flowers in upper whorls of inflorescence, × 1½ ; **4,** whorl of ♂ flowers with bract cut away, × 6 ;
5, ♀ flowers in lower whorls of inflorescence, × 1½ ; **6,** whorl of ♀ flowers with bract cut away, × 3 ; **7,**
achene, lateral view, × 6 ; **8,** transverse section of achene showing air-chambers, × 6. 1 from *Parker* ;
2–4 from *Lowe* 267 ; 5–8 from *Hancock* 7A.

from the petiole, 2–3 mm. wide. Peduncles shorter than the petioles ; inflorescence about 2 cm. long when young, with 5–10 whorls, the internodes expanding from below when older up to 5 cm. ; vegetative buds present, 1 or 2 of the outer bracts having a stiff, thickened midrib which becomes extended to 1–1·5 cm. long ; connate bracts 2–2·5 mm. long, usually with an entire rim, sometimes 3-lobed in the upper whorls. Male flowers: 4–8 in each whorl ; pedicels 2 mm. long ; bracteoles about 6, 1·5 mm. long ; sepals 3, spreading, broadly ovate, 2·5 mm. long, the two outer usually larger than the inner one ; petals 2 or 3, ovate, 1·5 mm. long ; stamens 3 ; filaments 0·5 mm. long ; anthers 1 mm. long ; abortive carpels 2–3, minute, globose. Female flowers : 3 in each whorl ; pedicels 3–4 mm. long ; sepals 3, ovate, 3–4 mm. long ; petals 1–3, minute, lanceolate ; staminodes 3, awl-shaped, minute ; carpels 3–6, ovate ; achenes 3–3·5 mm. long, smooth, slightly indented each side, air-chambers minute ; beak 1–1·5 mm. long. Fig. 6.

UGANDA.   Masaka District : Lake Nabugabo, Mar. 1935, *Hancock* 7A !
TANGANYIKA.   Kigoma District : Mlagarasi Swamps, Katare, 17.5 km. S. of Nguruka, 28 Aug. 1952, *Lowe in E.A.H.* 11581 !
DISTR.   U4 ; T4 ; Angola, southern tropical Africa and Madagascar.
HAB.   Soft mud and floating weeds in pools and shallow lakes ; 1000–1125 m.

The second African species, *W. schweinfurthii* Hook.f. (of which *W. sparganiifolia* Graebn. is a synonym), with a definite lamina to the leaf, ridged achenes, sessile ♀ flowers, and vegetative buds without hardened bracts, is distributed through West Africa to the Sudan and Northern Rhodesia.   Both these species are probably more frequent and widespread than they appear to be.

7. BURNATIA
Micheli in DC., Mon. Phan. 3 : 81 (1881)

Perennial glabrous dioecious, swamp and aquatic herbs. Leaves erect ; leaf-blade linear-lanceolate to ovate ; base decurrent to rounded. Inflorescence above the leaves, of whorls of 3 branches or 3 flowers ; bracts 3, lanceolate and membranous beneath the branches, much smaller beneath the flowers. Male flowers : sepals 3 ; petals 3 ; stamens 9 ; filaments flattened ; anthers oblong ; abortive carpels about 12. Female flowers : sepals 3, spreading ; petals 0, rarely present as 3 minute scales ; carpels numerous, free ; style ventral and very short ; stigma papillose ; ovules basal, solitary ; achenes bunched on the small stipitate receptacle, laterally compressed with a flange following the outline of the seed cavity on each side, glandular.

Monospecific ; known only from the African continent.

B. enneandra *Micheli* in DC., Mon. Phan. 3 : 81 (1881) ; F.T.A. 8 : 213 (1902) ; E.P. IV. 15 : 60 (1903). Type : Sudan, Kordofan, near Arashkol Mountain, *Kotschy* 192 (BM, K, iso. !)

Petioles up to 40 cm. long, usually about 30 cm. Leaf-blade acute, very variable in shape, from ovate with rounded base, to linear-lanceolate with decurrent base, e.g. from 13 × 7 cm. to 16 × 4 cm. and 14 × 1 cm. ; nerves 5 or 7, the median up to 1·5 mm. wide on the lower surface. Male inflorescence (15–)20(–30) cm., of 1–5 whorls of branches ; female inflorescence shorter ; lowest bracts 1–5 cm. long ; bracts and sepals whitish, sometimes tinged violet. Male flowers : bracts 1·5–2·5 mm. long ; pedicels 3–10 mm. long ; sepals erect, ovate, 2–3 mm. long ; petals smaller, delicate, persistent, about 1 mm. long ; stamens with filaments 1·5 mm. long ; anthers 1 mm. long ; abortive carpels oblong, compressed, 1–1·5 mm. long. Female flowers : bracts 1 mm. long ; sepals ovate, spreading, 1·5 mm. long ; petals, if present, 3, minute and scale-like, found only on very robust specimens ; carpels obovoid, compressed ; achenes 8–20, usually about 12, 1·5(–2·5) mm. long, flanges almost circular.   Fig. 7, p. 14.

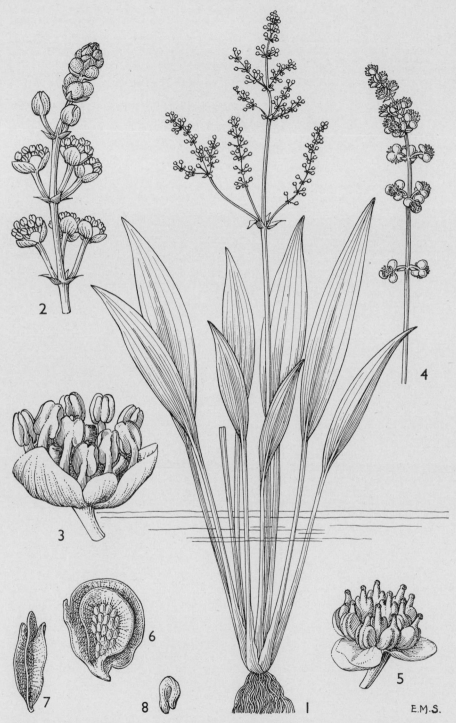

FIG. 7.   *BURNATIA ENNEANDRA*—**1**, ♂ plant, × ⅓ ;   **2**, part of ♂ inflorescence, × 2 ;   **3**, ♂ flower, × 9;
    4, part of ♀ inflorescence, × 2 ;   **5**, ♀ flower, × 9 ;   **6**, achene, lateral view, × 9 ;   **7**, achene, dorsal view,
    × 9 ;   **8**, seed, × 9. 1 from *Dalziel* 260 ; 2, 3 from *Rayner* 476 ; 4, 5 from *A. S. Thomas* 3558 ; 6, 7 from
    Bally 5237.

UGANDA. Lango District : Oruma, Moroto, Sept. 1935, *Eggeling* 2218! ; Ankole
District : Ruizi River, 18 Dec. 1950, *Jarrett* 255! ; Teso District : Amuria, 14 Sept.
1946, *A. S. Thomas* 4541!
KENYA. Northern Frontier District : Sololo, 3 Aug. 1952, *Gillett* 13685! ; Nairobi
District : Nairobi Game Park, 19 Aug. 1947, *Bally* 5237! & 8 km. N. of Nairobi,
8 June 1948, *Bogdan* 1637!
TANGANYIKA. Tanga District : flats by Lwengera River 5 km. E. of Korogwe, 16
July 1953, *Drummond & Hemsley* 3332! ; Magunga, 28 Dec. 1953, *Faulkner* 1316! ;
Moshi District : Mpololo, July 1928, *Haarer* 1498!
ZANZIBAR. Pemba Island : Vitongoge, 22 Sept. 1929, *Vaughan* 675! & *Mrs. Taylor*
84/2!
DISTR. U1–4 ; K1, 4 ; T2, 3 ; P ; widespread, from Ghana to the Sudan, and south-
wards to South Africa
HAB. Swamps, shallow lakes, and the edges of rivers and slow-flowing streams; 50–
1410 m.

SYN. *Echinodorus ? schinzii* Buchen. in Bull. Herb. Boiss. 4 : 413 (1896). Type :
South West Africa, Amboland, Ondonga, *Rautanen* 51 (Z, holo.)
Rautanenia schinzii (Buchen.) Buchen. in Bull. Herb. Boiss. 5 : 854 (1897) ;
F.T.A. 8 : 212 (1902) ; E.P. IV. 15 : 59 (1903)
Burnatia enneandra Micheli var. *linearis* Peter, F.D.O.-A. : 120 and Anhang
11 (1929). Type : Tanganyika, Uzaramo District, Dar es Salaam, *Peter*
45008 (B, holo.†)
B. alismatoïdes Peter, F.D.O.-A. : 120 and Anhang 11 (1929). Type : Tanga-
nyika, Uzaramo District, Dar es Salaam, *Peter* 45009 (B, holo. †)
B. alismatoïdes Peter var. *elliptica* Peter, F.D.O.-A. : 120 and Anhang 11
(1929). Type : Tanganyika, Pangani District, Pangani R., Hale, *Peter*
46584 (B, holo. †)
B. oblonga Peter, F.D.O.-A.: 120 and Anhang 11 (1929). Type : Tanganyika,
Uzaramo District, Dar es Salaam, *Peter* 46900 (B, holo. †)

INDEX TO ALISMATACEAE